Spinning Black Hole Inside Our Earth

by
Trillion Theory author
Ed Lukowich

Cosmology 5 book series (non-fiction) by Ed Lukowich
Trillion Theory (1) (cosmology) October 2015.
Trillion Years Universe Theory (2) (cosmology) 2014-2015.
Black Holes Built Our Cosmos (3) December 2015.
T Theory Says: Who Owns Our Universe (4) November 2016.
Spinning Black Hole Inside Our Earth (5) April 2017.
(Cosmology books published at Amazon).
See website: www.trillionist.com

Science Fiction novel authored by Ed Lukowich
The Trillionist, 2013 under pen name Sagan Jeffries.
Publisher: Edge Science Fiction & Fantasy Publishing.

Spinning Black Hole Inside Our Earth

Canadian Library ISBN: 978-0-9959663-0-7 (print)
ISBN Kindle eBook: 978-0-9959663-1-4
Library and Archives Canada Cataloguing in Publication
Lukowich, Ed, author **Spinning Black Hole Inside Our Earth**
(Trillion Theory Series) In print and electronic formats.
Canadian Library ISBN: 978-0-9959663-0-7 (soft cover)
ISBN Kindle eBook: 978-0-9959663-1-4
1. Black holes (Astronomy). 2. Earth (Planet) 3. Cosmology
1.Title.QB843.B55L84 2017 523.8'875 C2017-902720-4 C2017-903589-4

Table of Contents
Black Holes! Why do They Exist?

1: How Earth formed into a Sphere   page 01

2: How our Solar System Formed   page 10

3: How Galaxies Became Hotels   page 14

4: How Cosmos Grew So Enormous   page 22

5: Biggest Secrets of Black Holes   page 27

6: Basics of Trillion Theory (TT)   page 36

7: Cosmic Design is for Real   page 39

This 50-page book is a brief summary of (TT) Trillion Theory (T Theory). More in-depth is available in TT's other 4 books, via Amazon (print and eBooks) or (Inquire Chapters stores). **www.trillionist.com**

Many TT ideas may be very radically **alien** when first encountered. For, 'Most great ideas are initially viewed as ridiculous.' Read TT in full to grasp neo concepts, which when added together, make extraordinary sense.

T Theory predicts, 'TT's new theories are simply ahead of their time, by showing us what black holes really are.'

Astronomers and astrophysicists are presently moving towards discovering secrets which T Theory advocates. But their belief in the Big Bang slows their progress. That is why T Theory is now approaching the experts to provide credence to Trillion Theory.

Author *Spinning Black Hole Inside Our Earth.*
Founder *T Trillion Theory*

Luminary Canadian theorist Mr. Ed Lukowich (a Trillionist) presents a brand-new theory, uncovering major truths about hidden cosmic secrets which depict our cosmos to be a trillion years of age since its origin.

Spinning Black Hole Inside Our Earth, provides (in brief) new eye-opening answers to 4 great universe questions:

- How did our Earth form into a sphere?
- How did our Solar System get its formation?
- How did our Milky Way Galaxy become a hotel hosting millions of solar systems, such as ours?
- How did our cosmos evolve to its gigantic size?

Ed's answers to these 4 secrets shows that a specific type of scientific design was used throughout the project of building our cosmos; a design feature seen within atoms, planets, solar systems, and within the billions of galaxies. The notion of cosmic design does set off a firestorm with many people.

(Note: Detailed expanded answers can be found in Ed's 4 previous cosmology books. Also, note that the Ed Lukowich cosmology books only deal with the physical cosmos – these books do not deal with a spiritual side within our universe. For ideas on reincarnation, see Ed's futuristic novel *Trillionist,* under pen name Sagan Jeffries). www.trillionist.com

Now, following are answers to 4 great universe questions. Mysteries which Big Bang can neither unravel nor explain.

Chapter 1

How Earth formed into a Sphere

Do not believe in either a Big Bang or Nebular theory. These theories can't explain the fantastic detail found in atoms that occur on Earth and other cosmic spheres.

Then, how did Earth form an orb of matter comprise of atoms? Atoms which spin for eons (billions of years).

How Trillion Theory (T Theory) deals with our Earth

When examining Earth, here is what Trillion Theory (TT) **shows:** how atoms were formed; how Earth formed into a sphere; how atoms within Earth's core underwent extreme heat expansion; how this resulted in the first volcano millions of year ago. (A powerful volcano that buckled Earth's crust into continents and mountains).

Note: What TT does not deal with in its theory: how water, or vegetation, or animal life came to our planet. (Note: Earth, a Goldilocks planet in perfect proximity to our sun, was likely environmentalized and seeded by those who do that type of 'next step,' in bringing life to a hospitable planet capable of supporting life).

But first, let's back up some 5-8 billion years, to when Earth began to form as a sphere. **TT states that:**

TT states: Earth was formed by a powerful Black Hole

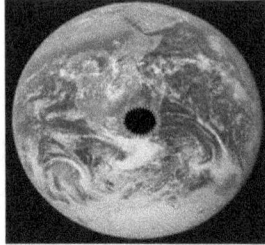

> The image shows a resident small size black hole spinning inside of Earth. The black hole's gravity carries ₐalong all our planet's matter and extends far past the surface to keep our moon in orbit.
>
> **T Theory says,** 'Prior to building Earth, this black hole spun ultra-fast as a naked black hole when billions of years in the past it attracted and spun light into matter, building a body to form Earth. Today, this cloaked black hole spins at a much slower rate because of its bulky acquired mass.'

'Whoa,' you say. 'A black hole formed our Earth?'

'Yes,' says T Theory. 'And, that black hole still resides at the central core of Earth.' Then, how about our sun?

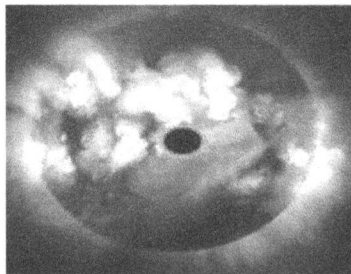

A cloaked black hole (XL size) resides inside of the core of our massive sun. The XL is losing control of its contents at a steady pace, but its strong gravity still holds our entire solar system in orbits.

T Theory says, 'Every spheroid, whether planet, moon of sun, was built surrounding a black hole. All of our solar system's 8 planets, 166 moons, and 1 sun are each controlled by a black hole residing deep inside them.'

In their building process, naked black holes used their powerful gravity pull to entice light, taking it past its known bending stage. Then, black holes deployed their talents to spin that light into atoms of matter.

Umpteen trillions of light rays disappeared for eons as they became imprisoned while spinning as atoms. Each black hole attracted and spun light until its entire inner, mid-body, and outer surface was plumb full. Although a totally full black hole could no longer spin any more light to matter, it did continue to attract and bend light to its surface, as does planet Earth. The rotational speed of each black hole slowed as it filled, yet its gravity still held objects on its surface and extended this gravity to other nearby moons.'

A black hole is responsible for building Earth? A black hole still resides inside of our Earth. Severe are such proclamations made by T Theory; enough to create a firestorm throughout the cosmology world.

Phase 1: Building a Sphere in T Theory

A naked empty black hole, with hundreds of spins per second, attracted and spun light into atoms of matter.

Phase 2: Many Types of Atoms Were Spun

The naked black hole spun tons of light into atoms as it first filled its inner helix and body compartments. Then, many more millions of tons filled its outer body with matter, cloaking the black hole from view. Light provided weight, giving the spheroid much mass.

Note: A large variety of elements (atoms) were spun dependent upon length and tenseness of light strands. This manufacturing process by the black hole was complex. (We see intricacies in subatomic particles). Black holes displayed exact preciseness, for spun light still spins unabated for eons, locked inside of atoms.

Phase 3: Controlling a Sphere

The earliest atoms spun into the inner core of a black hole 'came in hot' and were layered over by tons of incoming matter. Initially, this black hole maintained strong control over its hot inner atoms, but that control gradually waned as more layers piled up to its surface.

Phase 4: Loosening Always Occurs

Black holes always overeat. They always devour more light into atoms than they can properly control. Thus, the hottest atoms at the inner core of the black hole inside an orb always over-expand and then fight their way to the surface of the sphere via fissures.

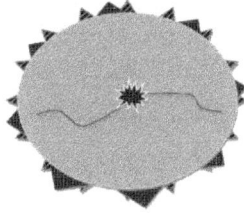

In some cases, this internal pressure can explode an orb spreading debris across space. We see many craters on orbs where meteorite debris impacted a surface.

On Earth (hundreds of millions of years ago) when the smallish black hole at the core of our planet lost control of its hot expanding inner fire, the pressure on fissures was so immense that Earth nearly exploded.

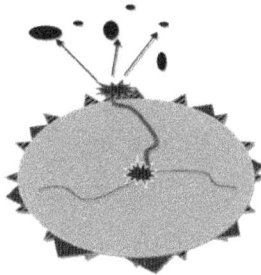

However, with Pimple Burst (Earth's inaugural titanic volcano), lava found a fissure to the surface, propelling rocks and debris into outer space. That volcanic release saved Earth from destruction as it quickly reduced the internal pressure. Even yet, initial pressure changed the surface: mountains were pushed up; continents moved apart; volcanic ash caused ice ages on the surface.

Our Sun Was Built a Little Differently

A sun evolves, through its approximate 15-billion-year cycle, faster than all of its planets and moons.

The sun became the central point for all the spheres in our solar system because it was built with the largest (XL) black hole at its core. (Note: Supermassive black holes at the hub of galaxies are even far larger than XL).

The sun's XL black hole size brings with it advantages and disadvantages (the good and the bad).

The naked XL black hole beat all of the other smaller sized naked black holes in a competition to attract and spin light into matter. Thus, it easily became the largest sphere of its group with the most mass and gravity; this enabled it to become the center of our solar system, holding all the other spheres in orbit. The bad, is that this gluttonous XL black hole ate the fastest, thereby it spun the loosest atoms. This meant an earlier internal fire for the XL, plus an early fire up of it entire body. The gravity pull from 176 other spheres continually tugged away at the sun's atoms. Today, more of the sun's atoms unspin and escape as rays of light.

The sun is the only fireball in our solar system. All of the lesser sized black holes at the core of planets and moons are presently in control of their internal fires.

Trillion Theory totally disagrees with astrophysicists, astronomers, and theorists such as a Stephen Hawking who for decades have been saying that light captured by a black hole was always trapped forever.

T Theory counters with: 'Light, trapped inside of a black hole will always escape back to light, even if it takes many billions of years to escape its imprisonment from lockup inside of an atom.' This is TT's recycling evolutionary law of the cosmos. Light is indestructible, no matter what, it will always escape imprisonment in the future and return to free traveling light.

Our sun is simply an XL black hole whose gluttonous eating habits led it to lose control of its atoms at an earlier stage than its competitors.

Law of Spheres by T Theory: 'All spheres, meaning all those with black holes at their core, will one day unravel like a sun and witness the unspinning of their atoms back to light. Suns culminate as a Supernova as they explode and lose total control. All of the planets and moons in a solar system undergo this unraveling back-to-light as they are melted when their sun goes Supernova; this is known as solar system Obliteration.

TT adds, 'A black hole can never be destroyed; it can survive a Supernova or solar system Obliteration.'

A cloaked black hole inside of an orb, once at loses all of its atoms, simply returns to being a naked black hole once again. At the end of a Supernova/Obliteration, the sun and all its other orbs unspin back to light. In the aftermath, the only evidence that remains of the solar system is the surviving graveyard of naked black holes. Note: At this point a black hole returns to its fast-naked spinning speed. An instantaneous whiplash causes the naked black hole to split in two in a replication process.

Replication fosters growth in sphere numbers, solar systems, and galaxies throughout cosmos. (Further T Theory detail is found in Ed's 4 book cosmology series). (Paperback or eBook, at Amazon).

Proofs for Trillion Theory (TT)

1st proof for TT might be found inside any sphere.

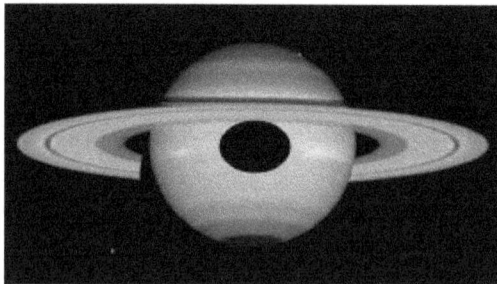

Scientists may discover a way to find a black hole inside of a sun, planet, or moon. That would prove T Theory. **2nd proof for TT might be found inside a lab.**

What if we make a miniature black hole in a lab? Suppose that we build a machine to spin light into atoms of matter inside a physics lab; this machine would emulate the actions of a black hole.

   T Theory calls this machine a Quantronix. An elongated neck pulls light strands into a chamber to be bent and then spun into atoms which are then spit out at the bottom of the machine.

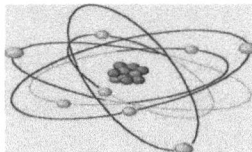

Chapter 2
How our Solar System Formed

How did our Solar System get its formation?
Once a person buys into the possibility of a black hole inside of each and every solar system sphere, then their orbital arrangement really begins to make sense.

T Theory looks back to the birth of our solar system. That scene began as a graveyard of naked black holes that had survived the Supernova, which exploded the sun of a prior solar system, and Obliteration of all orbs of that old solar system, on our very spot.

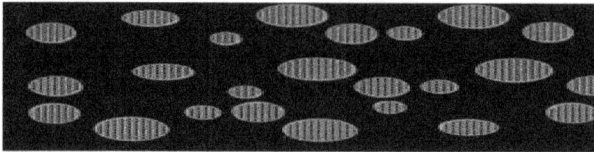

T Theory shows that our present solar system exists in the 67th of the 15-billion-year generations that occurred since the onset of our cosmos a trillion years ago. This ancient age far surpasses the ages of the current generation of stars in our sky which fool astronomers into dating cosmos at only 13.7 billion years. (TT shows that Supermassive black holes at hubs of ancient galaxies are 500-800 billion years old).

Our solar system began when all of the naked black holes, which survived the Supernova/Obliteration of the prior solar system, commenced a fight-for-light to build new spheres. (After destruction of the old solar system, Replication of the naked black holes meant a new larger solar system, or two new solar systems cast apart by the powerful Obliteration).

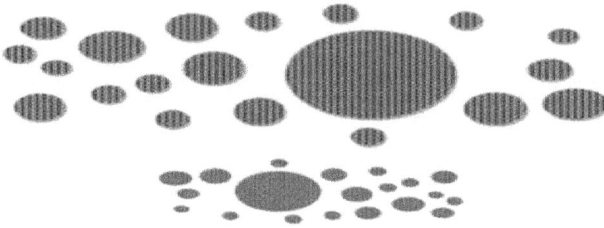

In the beginning of our solar system, the XL black hole from the sun of the old solar system had the greatest size and also shed its light (due to going Supernova) prior to any of the lesser sized black holes. Thus, XL had an upper hand on early light eating allowing it to be an even larger sun at the hub of our new solar system.

The solar system begins as the XL, growing into a sun, extends the greatest gravity to take all the other black holes (which are building their spheres) into orbits. At a distance from XL, some other large black holes utilize proximity gravity to hold smaller black holes as moons.

No two solar systems are ever exactly alike, as they differ in number of planets and moons. The formation of spheres follows the Black Hole Engagement Laws. 'Gravity of a larger black hole always overpowers lesser sized black holes in finalizing sphere arrangement.'

However, certain spheres will internally heat and go volcanic earlier than others. The smallest moons, which spun the tightest atoms, will be the last to ever heat up.

Whereas, the extreme XL black hole at the core of our sun, which spun the greatest number of loose atoms, was the first orb in our solar system to have an internal fire that raged so strong that its entire surface melted. One day, likely at a 10-14-billion-year age, our sun will totally lose control of its contents and go Supernova.

That will cause a subsequent Obliteration of this solar system. All that will survive is 177 naked black holes.

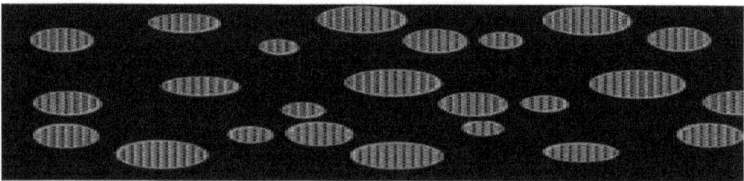

3rd TT proof would be found as a black hole graveyard. Astronomers may prove TT by finding a graveyard of naked black holes after a Supernova and Obliteration destroy a solar system. This is difficult to spot because of the profusion of light. After Supernova, appearing to implode, the naked XL black hole (from the old sun) devours tons of light released from the Obliteration.

4th proof for TT is found in the axial tilts of planets.

Earth's axial tilt on its axis is 23.44 degrees. Here's the rub. All the 8 planets orbiting our sun are in the same flat orbital plane, so Nebular Theory infers every planet, should have the same axial tilt when riding in orbit. But no. Neither Big Bang nor Nebular explains this mystery.

TT shows why each planet in a solar system has a unique tilt. Each black hole that spun a planet did its own axial tilt to gain its best possible strategic light attracting angle. That same tilt is still with the planet.

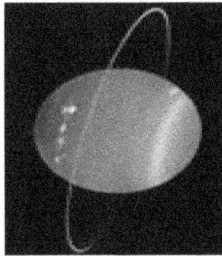

We view weirdo Uranus, spinning on its side. **TT says,** 'All black holes independently determine their axial tilt. Discovering how the black hole inside of Uranus got its bizarre sideways axial tilt is possible TT proof.'

Chapter 3

How Galaxies Became So Large

How did our Milky Way Galaxy become a host hotel for millions of suns and solar systems, such as ours?

T Theory portrays a whole new picture as to how galaxies came to be and how they grew so large.

Spiral galaxies are the most ominous sights in our cosmos. They show power, majesty, splendor, and magnificence as they flaunt their beauty. Galaxies can be described as gigantic remote islands in a space where cosmos boasts two hundred billion galaxies, most holding millions of solar systems.

While solar systems die on average every 15 billion years (or so), galaxies are permanent since they don't live and die by the 15-billion-year rule. Thus, solar systems recycle inside their galaxies, fostering growth so galaxies can get ever larger.

Spirals are 90% of the galaxies. Their bulging center has enough gravity to hold millions of solar systems in orbit. At this hub is a fast spinning supermassive black hole, nestled inside a clustered of stars. TT says, 'Inside of the bulge, the Laws of Black Holes are rewritten.'

'Think of a galaxy as a cosmic island hotel. The galaxy has millions of hotel guests, residing in solar systems. A gigantic galaxy easily survives a Supernova recycle of a single solar system, with the rest of the galaxy unaffected. While all of its solar systems get reborn about every 15 billion years and may appear to be 13.7 billion years old or less, look deeper to the supermassive black hole at the galaxy hub, which is up to 800 billion years old. The ancient supermasive black at the hub is the best recorder of cosmic history.'

How did a galaxy grow that old and that huge? Why is the black hole at the galaxy hub so supermassive? Great questions that T Theory is best suited to answer. 'Each galaxy began hundreds of billions of years ago as a single solar system, then grew gigantic into multiple solar systems around a master black hole.'

Milky Way Galaxy: from a small start, to gigantic today

Millions of solar systems in today's Milky Way. Follow TT's growth of our Milky Way, which started small.

♦ **800 billion years ago**, Milky began as a single solar system. The black hole centering its sun was XL size.

♦ **600 billion years ago**, Milky (via recycles of its solar systems) grew to hundreds of solar systems. The black hole at the center of the largest sun had grown and all the solar systems revolved around this XXL black hole.

♦ **400 billion years ago**, Milky now housed thousands of solar systems. The black hole at the central hub of the galaxy had grown into an XXXL sized black hole.

♦ **But then, catastrophe happened.** The XXXL black hole hubbing our Milky went Supernova. Obliteration struck at each solar system in the galaxy. This powerful cosmic act split the galaxy into two parts that hurled away from each other and reformed, far across space, as new galaxies. The XXXL black hole which reformed our new Milky Way determined to never again split its galaxy. So, it strategized how to become supermassive.

♦ **200 billion years ago,** Milky Way's black hole had grown to an XXXXXL size. It hubbed solar systems galore, as its moved towards being supermassive.

♦ **Today,** the black hole at the hub of our Milky Way hubs millions of starry solar systems. The black hole is now a permanent **supermassive** with a strategy of never having to go Supernova again and split its galaxy.

Note: Such terms seem befitting of an entity which is alive and able to make changes. TT Says, 'Yes, black holes are alive, programmed with goals to achieve.'

Black Hole Strategy: Massive to Supermassive

TT says, 'Think of cosmic evolution as survival of the fittest (best adapter). When black holes evolve, their goal is to control as much matter and as many orbs as possible. They strategize to use size advantage to out-duel lesser sized black holes in a battle for light so they can expand by overeating. The prolific masters of these feats are the ancient supermassives at hubs of gigantic spiral galaxies.'

When supermassive black holes, reach a certain benchmark size, they are able to evolve by the way they adapt to the eating of light. While still pulling in tons of light and devouring it, they no longer spin that light into matter for the purpose of building a sphere.

A supermassive deployed this smart strategy which allowed it to evolve much older and control its galaxy for longer than just one 15-billion-year cycle. It no longer had to someday go Supernova, destroying its galaxy. This way, the supermassive sustained itself for 100's of billions of years at the hub of a growing galaxy, while its occupant solar systems recycled and increased in numbers. The supermassive became a monarch, with an empire of suns and solar systems inside its galaxy.

The adaptation for the supermassive black hole meant making structural changes allowing it to deal, in a new way, with the tons of light which it pulled inwards.

In the image, we see the supermassive black hole at the hub of a galaxy. It is far different than what we see with a sun at the hub of a solar system.

♦ a supermassive black hole isn't cloaked inside tons of matter, like the XL black hole at the core of a sun.

♦ instead, the supermassive is surrounded, in its bulge, by a host of giant stars which it holds close to itself.

♦ the supermassive jettisons light out of its poles via plumes, indicating that it doesn't require that light.

♦ a supermassive still wants to attract light (stars in its bulge) but rids of itself of devoured light via plumes.

How did a supermassive black hole evolve and adapt?

T Theory says, 'Unless we are able to get super close to a naked black hole, or capture one, we might have to be satisfied with only guessing at its interior. T Theory makes that attempt here.'

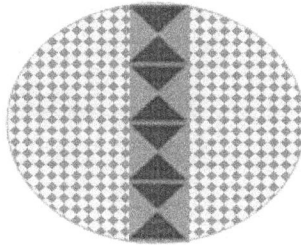

T Theory suggests: If we entered a naked black hole, we'd find billions of compartments, all ready to house atoms to be spun from light. Also, we'd experience a turbo-charged spin, the fastest in cosmos. At the center of the black hole, there is a spinning helix rod which deploys elasticity to lengthen and shorten as a means to perpetuate continual spin. This spin is so dramatic; a draft (gravity) extends out past the equator of the black hole, to pull inwards tons of light to be spun to atoms.

Inside a naked black hole, the spiral helix fills first, growing in length and girth. Then, more atoms fill the subdividing compartments to erect a fat body. Filling, stretching, and subdividing grows it larger. A motivated black hole wants to be the largest and most powerful.

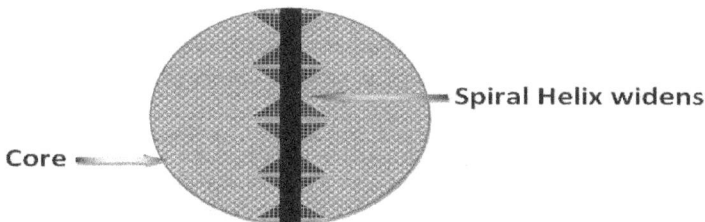

Core ◄——

Spiral Helix widens

The supermassive black hole makes structural changes in order to evolve. Such changes involved a widening of its spiral helix allowing it to become more hollow.

As the supermassive black hole ingested tons of light, it only spun enough light to matter to continually fill inner elastic compartments, thereby growing in size, without having to build a sun. The wide hollow helix became a conduit to pass light out through its polar exits. In essence, a breathing ability developed. A supermassive displays plumes of light jettisoning from its top and bottom. Thus, a supermassive strategized how best to center its galaxy for up to 800 billion years; destined to remain forever in that dominant position.

This evolutionary adaptation by a supermassive black means that it no longer has to grow into a sun. Now, it can live long to control its galaxy. Also, it can spin fast because it stays as a black hole, never taking on a sun's sluggish rotation. Yet, it can still pull huge stars into clusters around itself, ingesting their light. Then, it can spew and jettison that light back out as plumes. (It can use the mass of these clustered stars around itself to assist with its mass to control an entire gigantic galaxy).

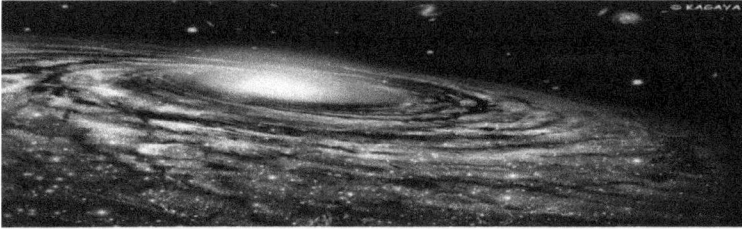

Clearly seeing a galaxy's supermassive black hole isn't easy due to a brilliant halo from a plethora of these stars in concentration surrounding the supermassive black hole. T Theory says, 'Strategy by the supermassive allows it to pull dominant stars into a cluster around itself. Those stars are slowly gobbled up and never permitted a chance to challenge for the supremacy of the galaxy. With ingenious strategy, a supermassive uses its prowess to make the stars subservient in aiding it to control the vast humongous galaxy.

5th TT proof found in a supermassive black hole.

Should astronomers be able to delve deeper into the age of a supermassive black hole at the hub of a galaxy, they may discover the supermassive to be hundreds of billions of years ancient; far older than the 13.7 billion year present presumed estimate for the age of cosmos. GROWTH + EVOLUTION replace a Big Bang origin.

No to Big Bang

Chapter 4
How Cosmos Grew Enormous

How did our entire cosmos evolve to its ginormous size? **Incorrectly, Big Bang proclaims** that an explosion of matter into space, followed by Nebular action, formed all of the spheres, solar systems, and galaxies of our cosmos. All of the matter available to our cosmos occurred with the Big Bang. New matter cannot be created to grow the cosmos larger.

T Theory says, 'Wrong. **Quit thinking in Big Bang terms.**

> T Theory says false to the Big Bang theory for:
> • estimating the cosmic age at only 13.7 billion years by incorrectly using just current stars.
> • depicting cosmic origin as coming from an explosion of matter expanding outwards.
> • concluding that galaxies receding from each other proved an explosive origin to cosmos.
> • accepting a Nebular theory which incorrectly proposes that solar systems/galaxies are the result of nebular clouds swirling and cooling.

For, TT shows that cosmos has an inexhaustible supply of energy available for black holes to continually build more spheres and solar systems in growing galaxies.

For, space didn't yet exist at the cosmic origin. Over a trillion years, as black holes ate from an ocean of light to build the cosmic spheres, the vast depleted cavern became what we today see as void, empty, black, weightless, space.

<u>Trillion Theory: Cosmos began from an ocean of light</u>

Cosmos is tricky; often it fools our eyes and reasoning.
We tend to think that since the contents of our cosmos
sit in space, that space was there first as the box which
cosmos was born into. (In Big Bang, supposedly matter
exploded into space, a space which was already there
as an empty area, ready and waiting to accept matter.

  But, T Theory says, 'NO! Space was never waiting with
open arms for cosmos to occur. TT states, 'Space was
not first. Space is just a huge vacated area caused when
black holes spun vast amounts of light into spheres.'

  TT says, 'Picture an earliest cosmos as tons of light
(along with its properties such as weight) spun by black
holes into ultra-tight sphere packs. Once this task was
done, all light was inside of the spheres, while space
was a vast, empty, void, weightless cavern.'

  <u>Much savvy was needed to start a universe</u>
'The scientific strategist who began our cosmos started
with bare tools; those tools were light and black holes.

TT says, 'At a cosmic origin, a trillion years ago, all
that existed was an endless ocean of frozen light.'

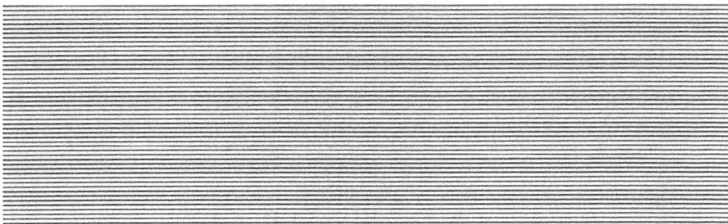

Frozen light was foreign to us. Then, in 2014, physicists froze a light ray in their lab, slowing its speed to zero.

'Today, if we traversed to the edge of space. we'd come to the outer perimeter of the space cavern. We'd smack into the hard-surfaced walls of the frozen ocean of light surrounding a long cucumber shaped cosmos.

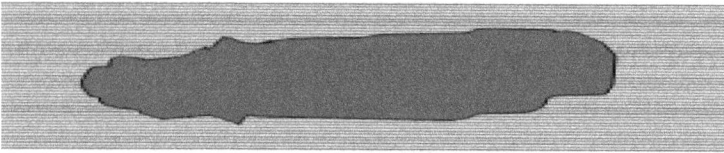

The outer ocean holds an endless supply of light to be resourced by naked black holes and spun into even more cosmic spheres. At the ocean walls, we'd find new worker-bee naked black holes, busy breaking off chunks of light, freeing that light to supply a cosmos which is forever growing in spheres, solar systems and galaxies.

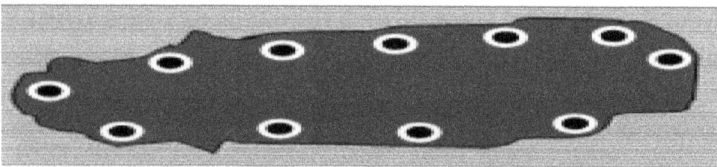

The black holes tightly spun light from the ocean into compressed spheres around themselves. A vast cavern of empty space resulted inside the frozen light ocean.

T Theory says this about the origin of our cosmos. 'The opening cosmic event began a trillion years ago, with the interjection of the first naked black hole into the endless ocean of static frozen light. To free light for its eating, the spinning black hole broke chunks from the ocean wall, spinning and capturing light into atoms of matter round itself, as it built the first cosmic sphere. The large eaten-away area was replaced by an empty cold weightless void which TT calls the space cavern.'

TT says, 'An immense amount of light is needed to supply the atoms for a body of matter around a sphere. Formula $E=mc^2$ calculates this staggering ratio of ocean light (zeptillion tons) that goes into the volume of a sphere. Space takes 99.9% of the cosmic area; whereas, galaxies, solar systems, and spheroids take up less than .0001%. Conversely, one supernova star releases megatons of energy.'

So, when I look into Earth's evening sky, I see spheres (moons, planets, stars) tightly compressed from the spinning of megatons of light. Around the spheres I see a vast emptiness of space which is now void of matter.

How Cosmos Grew Ever Larger

Cosmos used, and still deploys, many ways to grow larger: more black holes introduced into the light ocean continually built more spheres; supernovae double the number of solar systems; massive black holes can grow solar systems into galaxies; galaxies going supernova results in two new receding galaxies, flung apart.

T Theory estimates that cosmos doubles in size (number of stars and solar systems) and adds some new starter galaxies with each new 15-billion-year generation cycle. Astronomers now estimate 74 quintillion stars inside billions of galaxies.

TT says, 'The numbers will double to 148 quintillion cosmic stars in the next 15 billion years. Each minute, another star goes supernova somewhere in the cosmos, thereby setting in motion conditions for the surviving black holes to build the cosmos ever larger, while expanding space ever wider.

Proclamation from T (Trillion) Theory (TT).

Small changes give rise to large consequences. Astronomers will need to undertake a far greater cosmic introspection by performing forensics on black holes, determined to find all of the truths.

Chapter 5

Biggest Secrets of Black Holes

Albert Einstein laid a theoretical basis for the existence of black holes by predicting that light bends when it nears a sphere's gravity.

Yet years later, we are only beginning to know black holes. Till T Theory, no one had ever surmised that a black hole resides for billions of years at the core of every cosmic sphere. Till TT, no one had any inkling to relate black holes to the building of our cosmos.

T Theory says, 'Black holes are *sphere factories* hard at work to build and then recycle cosmic spheres. Also, via their gravity they provide the organization to control all spheres, solar systems and galaxies.'

A Black Hole is not really a hole, but rather on object which has super density, with massive gravitational pull.

Weirdly mysterious. Quite simply, what we don't yet know about black holes far outweighs anything we do. Prior to TT, no one was able to explain their presence. Black holes had always been portrayed as monsters. Finally, some astronomers are converging to TT, seeing how black holes bring orderly cosmic organization.

Penn State astrophysicist Yuexing Li had this to say, 'Now we see that black holes were essential in creating the universe's modern structure.'

Here is what astronomers previously thought:
- black holes were a rare cosmic feature.
- black holes brought only chaotic destruction.
- black holes sole purpose was to rip apart a star.
- black holes were a one-way street for light.
- black holes all spun in the same direction.

Here is what many astronomers still <u>incorrectly</u> think: 'Black holes are quicksand, trapping light forever.'
TT replies, 'Astronomers/astrophysicists have difficultly placing the existence of black holes into Big Bang.'

So, to this day, black holes are very misunderstood.

Here are new TT ideas which some astronomers are now discovering about black holes:
- black holes are a common cosmic feature.
- black holes are becoming easier to find.
- black holes extraordinarily exist by the billions.
- black holes aren't just chaos and destruction.
- black holes can spin at millions of mph.
- black holes might spin in either direction.
- black holes also have another gentler side.
- black holes are purposeful cosmic organizers.
- black holes seem to shape our cosmos.
- supermassive black holes are at galaxy hubs.
- a supermassive controls and organizes a galaxy.

Yet, T Theory says, 'Keep looking much deeper.'

TT states, 'Look for black holes in these settings:

▶ cloaked within cosmic spheres, hidden from view, no longer naked and empty, but brim-full with matter.

▶ on the perimeter of space, at the ocean of light, whereas worker black holes they break off chunks from the frozen ocean and devour freed light in matter.

▶ in a graveyard after a Supernova, where all the black holes that survived Obliteration of the old solar system are Replicating and ramping up for battle against their competitors in building spheres for a new solar system.

▶ at the hub or galaxies as supermassive black holes.

Here are more TT ideas about black holes:
- A black hole eats light spinning it into matter.
- A black hole spins light into a body of matter.
- A black hole (BH) is a cosmic sphere builder.
- A cloaked BH resides inside of every sphere.
- BH spheres are planets, moons, and suns.
- BH resides up to 15 billion yrs inside a sphere.
- A BH sheds its sphere at its cycle's end.
- A black hole is totally indestructible.
- A sun's black hole can survive a Supernova.
- BHs survive Obliteration of their solar system.
- A BH always survives the death of its sphere.
- A BH splits (replicates) at its survival moment.
- An exposed BH always rebuilds a new sphere.
- A black hole can build 1 sphere in each cycle.
- A black hole can build a sphere in every cycle.
- Black holes populate their solar systems.
- Solar systems recycle within their galaxies.
- Solar system numbers increase inside galaxies.
- Ancient galaxies grow in numbers and in size.
- Recycling was (is) on-going for a trillion years.

T Theory says, 'One secret which astronomers have yet to discover is that all black holes are sphere factories, spinning light into orbs. Black holes then reside inside of the spheres for billions of years. Therein, as spinners of light to matter, black holes built the orbs in all solar systems. Indestructible black holes have been the sphere masons and recyclers for a trillion years of cosmic history.'

TT says, 'Black holes are ubiquitous to our universe. T Theory predicts that black holes will be discovered existing all across cosmos. Already, we know of the supermassive omnipresent black hole at a galaxy hub. TT predicts that an x-large (or massive) black hole is at the core of every sun; and a smaller black hole at the core of every smaller sphere. TT says, 'These black holes provide the spin and gravity common to all spheres.'

TT, 'The even bigger questions are: how did black holes become cosmic builders? Where are they from? Who owns this Black Hole Society? Or are black holes the active owners of our mysterious universe?

Now, T Theory takes on Stephen Hawking, who is a fantastic spokesman for cosmology. However, he does sometimes get a few ideas wrong.

Hawking, the renowned English theorist/cosmologist, in April of 2013 admitted to a rather large blunder. Till then, he'd thought that light swallowed up by any black hole was seemingly lost forever. He has now recanted, admitting radiation escapes. Hawking adds, 'Black holes exist differently than first thought;' adding that while light can't escape from a black hole, the light is sort of stuck or stored in a holding pattern inside a black hole.

T Theory likes that Hawking is finally seeing the light. TT says, 'Black holes spin light into matter. But matter always returns to escaping light, even if it takes billions of years. A sun is an example where a tired black hole has lost the grip which it held on its matter. Each day, billions of atoms escape as light from a sun; as the sun further ages, it is destined to someday go Supernova. Light will always escape a black hole and return back to free traveling light, even if it takes billions of years.

Till T Theory, no one ever rationalized that black holes have a true higher purpose as ultimate cosmic builders.

Till T Theory, no one ever surmised that light escapes a black hole after an entrapment of billions of years.

Till T Theory, no one theorized how eventual escape by light from a black hole (or a sun) is an integral part of the recycling process of the spheres of our cosmos.

Till T Theory, no one knew of a black hole's goal is to be the strongest player, to dominate lesser black holes.

Till T Theory, no one postulated a black hole living at the core of every cosmic planet, moon, and sun; and the was responsible for having built that sphere.

Till T Theory, no one ever realized that the reason Earth rotates on its axis and extends **gravity** outwards is truly credited to the black hole residing inside of Earth.

The shape of the gravity projected by a black hole is in a disc-like plane out past its equator. This flat gravity field is prevalent around all naked black holes; around all black holes inside suns, planets, and moons; and around all supermassive black holes at the hub of spiral galaxies. (A spiral galaxy shows its flat gravity shape).

'Gravity is the super-glue projected by black holes. When naked, a black hole turns its fast spin into gravity to attract and spin light into matter. After getting full, a cloaked black hole within a sphere spins slower using its gravity to keep contents on the surface and to hold other spheres in orbit. Thus, gravity is also the sheriff which enforces the cosmic Rules of Engagement between all of the black holes within solar systems and galaxies.'

TT, 'Gravity from black holes has definite functions:'

Gravity is the strong gravitational pull which an ultra-fast spinning NAKED black hole extends outside of its core to attract and then consuming light into matter.

Gravity is the strong pull which a CLOAKED black hole extends to and out past the surface of the sphere which it built around itself when it spun light into matter. This gravity can hold lesser sized spheres in orbit.

Gravity is the enormous pull which a SUPERMASSIVE black hole extends out to all of the suns and the solar systems inhabiting the galaxy which it controls.

Gravity throughout cosmos is supplied and OWNED by all the black holes in control of our cosmos.

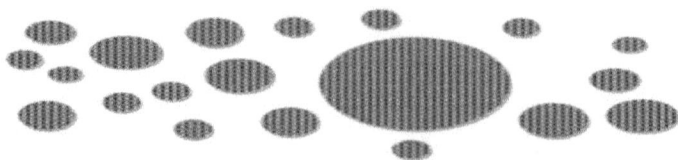

TT's black holes help answer cosmic puzzles, such as:

1: The 13.7-billion-year age for cosmos was jeopardized

in 2014 when astronomers discovered the 15.5-billion-year-old Methuselah Star. How could a star out-age the cosmos? A paradox and a fly in their ointment.

TT readily shows that the majority of stars in our current sky belong to our present generation with ages ranging a few billion to 13.7 billion years. But, since black holes recycle the stars, that recycle is different for each star. A cosmic rule in a trillion-year cosmos is that the upper limit for a black hole to occupy a star prior to it going supernova is 15-16 billion years.

2: Why does a supernova sun implode back into itself?

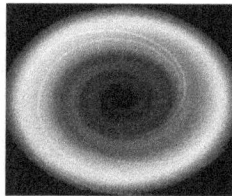

TT: Astronomers propose that black holes comes from a supernova sun which implodes back into a black hole, which they call an ultra-dense dwarf or neutron star.

But TT credits the black hole as the builder of that sun in the first place. When supernova occurs, all matter is exploded away, and we are left with the original naked black hole which built that sun. It appears to collapse because the black hole is immediately devouring new light densely into its bowels via its prolific gravity.

Chapter 6
Basics of Trillion Theory

T Theory asserts the following basics:

- Cosmos, at a trillion years, is far older than the 13.7 billion years estimated by astronomers.
- Cosmos grew 73 quintillion stars (No Big Bang).
- Black holes are cosmic builders (machine-like) which spin light into matter to form spheres.
- A black hole (sizes vary) resides at the center of every moon, planet, star, and at a galaxy hub.
- The full black hole residing inside of Earth gives our planet its inherent spin and its gravity.
- Solar systems, and all their spheroids, comprise the units inside the galaxies of our cosmos.
- Galaxies have permanency, while resident solar systems recycle every 15 billion years, or so.
- Each recycle of a solar system ups the count of stars and solar systems inside ancient galaxies.
- There's an endless energy supply (light ocean) which black holes access to grow the cosmos.
- Light, with its incredible properties, is deployed as a singular material in the recycling process.

How Our Cosmos Grew Large

- Cosmos had 66 cycles of 15 billion years.
- We are in cycle 67, in year 1 trillion.
- Our Cycle 67 has over 73 quintillion stars inside of billions of solar systems and galaxies.
- Cycle 68 will have over 146 quintillion stars.
- One generation cycle is typically 15 billion years.
- Black holes do recycling, growing cosmos larger.
- A sun's black hole going Supernova is a key.
- Black holes are indestructible.
- Black holes survive Obliteration of a solar system.
- Black holes always start up a new solar system.
- Black holes replicate (split) after a Supernova and a subsequent Obliteration of their solar system.
- Black holes, which replicate, then build new solar systems to continually grow cosmos ever larger.
- Solar system numbers increase inside galaxies.
- Ancient galaxies grow in numbers and in size.
- Recycling was (is) on-going for a trillion years.

Today, inside an endless light ocean, cosmos has grown to be a trillion light years across a space cavern. Tightly compressed is all the matter taken from the light ocean and spun by black holes into spheres inside of galaxies.

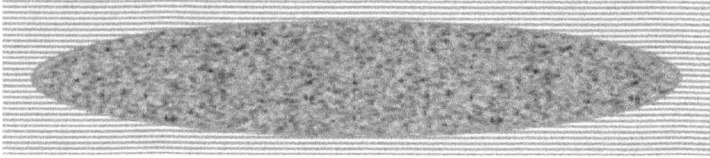

Cosmos (space) is a linear shape. Black holes, building all solar systems and galaxies, have eaten light from the outer light ocean along a flat type equatorial place.

TT says, 'During the trillion-year cosmic evolution, it is impossible to overstate the vital role of black holes as sphere builders, and the impact that the Supernovae of suns have had on cosmic growth and recycling.'

Because of the phenomenal spinning powers of black holes, and Supernovae of suns (which they did build), cosmos perpetuates its own growth. 'Nobody has to run out to turn a crank or plug-in a charger when it is time for cosmos to recycle a solar system in a galaxy.'

Chapter 7
Cosmic Design For Real

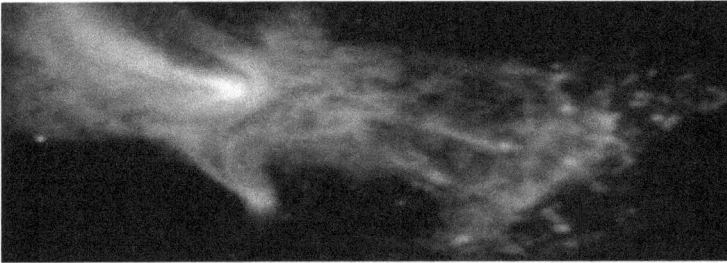

T Theory alleges that a scientific hand was behind the design of our cosmos. This is seen by the deployment of just one basic material, namely light, in construction of all the cosmic spheres. Then, equally great scientific genius is displayed in the crafty design of black holes, enabling them to be the cosmic builders and operators. Black holes are so strategically clever. This is shown by the manner in which they supply a method to recycle, grow, and perpetuate our cosmos from one generation to the next larger one. Cosmos is amazing engineering.

Cosmos always pulls us towards the next step:

In 2004, a Gemini Telescope found galaxies more fully mature than one might expect in a cosmos of only 13.7 billion years. **Dr. Robert Abraham, Dept of Astronomy, U. of T.** 'We are seeing that a large fraction of cosmic stars were in place when the universe was young, that shouldn't be the case. This glimpse back in time shows pretty clearly that we need to re-think what happened.'

Dr. Patrick McCarthy, Carnegie Inst. also added, 'It's unclear if we need to tweak the existing models or develop a new one to understand this finding.'

Many prominent people agree. 'Astronomers admit that they must re-think our universe, since numerous large cracks have appeared in Big Bang theory.' A Bang explosion origin fails to explain the workmanship that went into cosmic complexities. Astronomers have erred in estimating cosmos to be only 13.7 billion years old. As well, Big Bang and Nebular theory are far off the mark in explaining how our cosmos truly formed.

TT says, 'A magnificent scientific design is at work in our cosmos. Superlatives are simply inadequate to describe the scientific genius behind cosmos. If it were a contest, our universe wins in a runaway.
Now, it is imperative for astrophysicists and for astronomers to agree on a new paradigm model.'

Theory of Everything
ToE

'Whether we're looking inside of an atom, or viewing a gigantic galaxy, we need to perceive all cosmic features from a strategic point of view. Black holes always do things for a reason; wanting to be the biggest and in control of the most by over-powering or out-foxing.'

Here is such a strategic instance: An XL black hole at a sun's core rids itself of its contents before any other orbs in its solar system. Thus, XL gets naked and is first in line to eat light when Supernova and Obliteration hits the solar system. Therein, XL has the upper hand to grow larger and be in charge of the new solar system.

T Theory says, 'Brand new theory vastly expands our minds. No great idea was ever immediately accepted; most are initially viewed as ridiculous.'

The GOALS of Trillion Theory (TT) are genuine

- **Provide a succinct accurate theory** explaining cosmic origin and functions from a scientific design approach.
- **Show cosmic age at a trillion years** since origin.
- **Sell the notion that our cosmos began small** and then grew prodigiously via its 15-billion-year generations.
- **Credit black holes**, as scientifically designed machine-like entities for the growth in numbers of solar systems and galaxies throughout the cosmos.
- **Displace the Big Bang and Nebular paradigms** which have been wrongly accepted.
- **Show Trillion Theory as correct theory**, demonstrating its answers to: age, origin, and operations of our planet, solar system, galaxy, and the entire cosmos.
- **Indicate TT proofs** and suggest ways astrophysicists and astronomers can assist to validate these proofs.

No other theorist has ever made as neo a theory as T Theory which approaches our cosmos from a designer's strategic perspective based upon the best which ultra-science could ever possibly offer.

Carl Sagan said, 'It is better to grasp the universe just as it really is than to continually persist in delusion, however satisfying and reassuring.'

In Conclusion

Trillion Theory (TT) is a new brash cosmology theory proposing many new radical ideas about our universe. These ideas are in brief form in this book. They can be viewed in full detail in Ed's **Cosmology 5 book series:**

Trillion Theory (1) (cosmology) October 2015.

Trillion Years Universe Theory (2) (cosmology) 2014-2015.

Black Holes Built Our Cosmos (3) December 2015.

T Theory Says: Who Owns Our Universe (4) November 2016.

Spinning Black Hole Inside Our Earth (5) April 2017.

(Cosmology books published at Amazon).

See website: www.trillionist.com

Author's Disclaimer

The Trillion Theory theories in this book were developed solely by the author Ed Lukowich. No other person may claim to be the originator, founder, or author. The material is protected by copyright; any reproduction or distribution cannot take place without the author's consent.

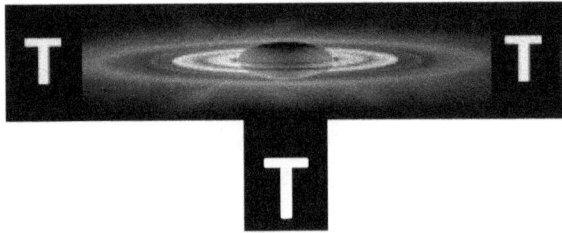

'If you like T Theory, please find a way to support it.' Thanks for reading.

Ed Lukowich

www.ingramcontent.com/pod-product-compliance
Lightning Source LLC
Chambersburg PA
CBHW022054190326
41520CB00008B/786